Islands of the Mediterranean
in pictures

VISUAL GEOGRAPHY SERIES

A craggy isle rising out of the waters of the Bay of Naples, Capri has been praised in song and verse since Roman days. Once the unofficial capital of the Roman Empire, the island has always been prosperous, perhaps more so than any other island in the Mediterranean. The climate here is invigorating and there are an endless number of spectacular vistas from the jagged highlands.

By COLEMAN LOLLAR

 STERLING PUBLISHING CO., INC.   NEW YORK

Oak Tree Press Co., Ltd. London & Sydney

VISUAL GEOGRAPHY SERIES

Afghanistan
Alaska
Argentina
Australia
Austria
Belgium and Luxembourg
Berlin—East and West
Brazil
Bulgaria
Canada
The Caribbean (English-
 Speaking Islands)
Ceylon
Chile
Colombia
Czechoslovakia
Denmark
Ecuador
England
Ethiopia

Finland
France
French Canada
Ghana
Greece
Guatemala
Hawaii
Holland
Honduras
Hong Kong
Hungary
Iceland
India
Indonesia
Iran
Iraq
Ireland
Islands of the
 Mediterranean
Israel

Italy
Jamaica
Japan
Kenya
Korea
Kuwait
Lebanon
Liberia
Malaysia and Singapore
Mexico
Morocco
New Zealand
Norway
Pakistan
Panama and the Canal
 Zone
Peru
The Philippines
Poland

Portugal
Puerto Rico
Rumania
Russia
Scotland
South Africa
Spain
Surinam
Sweden
Switzerland
Tahiti and the
 French Islands of
 the Pacific
Taiwan
Thailand
Turkey
Venezuela
Wales
West Germany
Yugoslavia

PICTURE CREDITS

The author and publishers wish to thank the following for the use of the photographs in this book: Embassy of Spain, Washington, D.C.; French Embassy Press and Information Division, New York; French Government Tourist Office; Greek National Tourist Office; Italian Cultural Institute; Italian State Tourist Office; KLM Royal Dutch Airlines; Spanish National Tourist Department, New York; Trans World Airways; United Nations, New York; Vilko Zuber, Yugoslavia.

The citadel of Bonifacio, Corsica looms above the quiet waters of the port.

CONTENTS

1638417

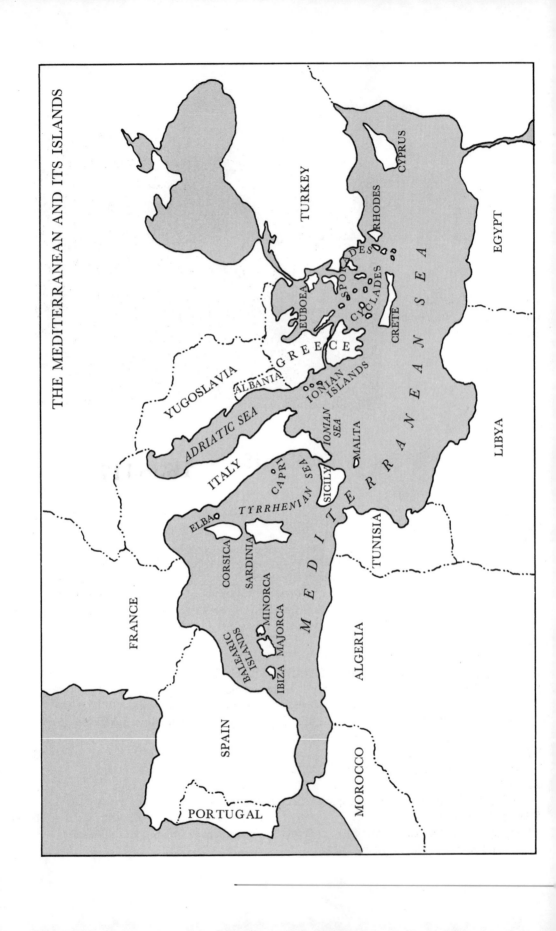

THE MEDITERRANEAN AND ITS ISLANDS

FRANCE

YUGOSLAVIA

TURKEY

SPAIN

PORTUGAL

ITALY

ALBANIA

GREECE

EUBOEA

SPORADES

RHODES

CYPRUS

CYCLADES

CRETE

ADRIATIC SEA

IONIAN ISLANDS

IONIAN SEA

MALTA

SICILY

CAPRI

TYRRHENIAN SEA

ELBA

CORSICA

SARDINIA

MINORCA

MAJORCA

IBIZA

BALEARIC ISLANDS

MEDITERRANEAN SEA

TUNISIA

ALGERIA

MOROCCO

LIBYA

EGYPT

INTRODUCTION

MOST OF THE islands of the Mediterranean belong to Greece, Italy, Spain, France and Yugoslavia, while two of them, Malta and Cyprus, are independent countries. Politics aside, however, these sunny islands, strung across the sea from the shores of Spain to the Turkish coast, have much in common.

First of all, they have the same deep blue sea around them and bright blue sky above. Most of the islands are the tops of fold mountains, created by the same pressures that gave birth to the Alps and Atlas mountains and the great depression that became the Mediterranean Sea. A few are volcanic in origin.

The Mediterranean type of climate, said to be the most pleasant on earth, is found in just a few other places, including Southern California, Western Australia, and the tip of South Africa. Though temperatures and rainfall may vary slightly from island to island because of altitudes and prevailing winds, it seldom gets extremely hot or cold anywhere. The summers are warm and dry, the winters mild and damp.

The history of the Mediterranean islands is, to a great extent, the story of Western man. Through much of history, they were at the crossroads of the known world.

Most of the islands were touched by the succession of empires that spread across the Mediterranean Sea. The Minoans, first to set up island outposts, were themselves from the island of Crete. The Phoenicians sailed from the coast of Lebanon and Syria between 1250 and 774 B.C. to set up colonies on many islands. Then the Assyrians, during their period of greatness between 910 and 606 B.C., invaded Cyprus. The Etruscans ventured from their home on the Italian peninsula to nearby islands between 900 and 400 B.C.

Greece was the first of the great powers, setting up its first island colonies about 850 B.C. and ultimately controlling the Mediterranean, challenged only by Carthage, North Africa's greatest empire. Rome came later, taking root in the Mediterranean about 300 B.C. After the Punic Wars, when Carthage was driven from Sicily, Rome became the only power ever to rule all the Mediterranean islands at one time.

The Goths—those "barbarians" from Central Europe who destroyed Rome— spread throughout Europe and took several of the islands, but they declined quickly. Islam, through the Moors in the western Mediterranean and the Turks in the east, held practically every island at one time or another during the Middle Ages. The Christians—Byzantines, Normans, the Papacy, the Italian city-states and the Holy Roman Empire—often fought with each other for the choicest of the islands, when they were not battling the Moors and Turks.

Today these Mediterranean islands are beloved by people in search of Europe as it used to be, for they are still very much the Old World. Not one of them is famous for its industrial development or business activity. Instead, simple people still farm the land—olives, grapes, and fruits being the most prevalent crops—and catch endless varieties of fish in the still unpolluted deep blue waters.

The Cuevas de Arta, some of the world's most impressive stalactite caves, are found beneath the coastal hills of Majorca.

THE GREEK ISLANDS

The isles of Greece have been as important to the history and development of the country as any area on the mainland. Greek civilization, in fact, began on the island of Crete. Thanks to the great number of islands with good ports around its shores, Greece has always been a great maritime power. The olive oil, wines, and other products of the islands contribute to the nation's strength as a trading power. And for a long time, the incomparable beauty of these islands has lured increasing numbers of foreign visitors to Greece.

CRETE

Among the Greek islands, Crete claims the most superlatives. It is the largest (160 miles long and as wide as 35 miles), has the longest and grandest history, and probably is the most important economically today.

Crete lies about 150 miles southeast of Athens, between the two stretches of the Mediterranean called the Aegean and Libyan Seas. It is a long, narrow strip of land with groups of mountains that divide the island into four natural regions.

The island's four provinces correspond to these regions.

The northern coast is rocky and indented, and as one might expect, a large number of coastal villages and seaports are found there. The south coast is smooth, with no natural ports and in some places the mountains form rock walls against the sea. In the very heart of Crete, Mount Ida, the legendary birthplace of Zeus, rises above the plains.

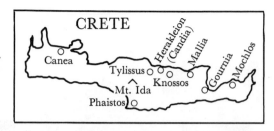

The interior is topped with numerous plateaux and basins, ideal for pastures. Once the land was covered with great pine forests but these have mostly disappeared. The occasional carobs, strawberry trees, wild olives and junipers add interest and aroma to the flora.

The largest city on Crete is Herakleion, fifth largest in Greece. Here is the spectacular Archaeological Museum, whose 23 halls contain perhaps the greatest collection of ancient Mediterranean artifacts. Most of the island's more important diggings are located near the city, especially at Knossos, 3 miles southeast, where the palace of Minos was uncovered.

Canea, though smaller than Herakleion, is the capital. It is a busy town, with many modern buildings and broad avenues that end at the sea.

The best modern example of what an ancient Minoan city was like is Gournia, a seacoast village (though it may have not been so near the sea in ancient times.) Houses here are still built of brick and stone just as they were thousands of years ago.

At the small village of Phodele, the great artist El Greco was born, but none of his works are in Crete, since he made his home in Spain.

History on Crete dates back to 5000 B.C. with the establishment of the splendid Minoan civilization, which was contemporary with the Egypt of the Pharaohs. This was the Mediterranean's first island civilization, and though forgotten for most of history—or accepted only as a legend—the Minoans had a profound influence on the development of Greece, Rome, and all of Europe.

The great cities of ancient Crete, the remains of which are now the island's principal attractions for tourists and scholars, were built in two separate periods and each ended in almost total destruction. The first of the great palaces was built about 2000 B.C. and lasted until 1700 B.C.

Discovery of the Palace of King Minos at Knossos on Crete at the end of the last century proved the existence of the great Minoan Civilization. This elaborate place has many well-preserved frescos and a plumbing system that could still work!

The town of Hydra is known for its vivid buildings clinging to the hills. The port below the town has a long seafaring tradition.

The second period, principally famous for mansions and private estates, began 100 years later and abruptly ended about 1520 B.C. No reason is known for either period's rapid downfall, though a natural disaster is suspected for the second. However, enough of the writings and art of the Minoans have survived to tell a fairly complete story of ancient Crete.

The Cretans, it seems, were similar to the Iberians who inhabited Spain and the Balearic Islands at the opposite end of the Mediterranean. Cretan kings were always called Minos—and this accounts for the name of the civilization. They lived in grand palaces with advanced conveniences, such as running water and bathrooms. Frescoes record the daily life of the people as filled with pleasures and sports, including an early form of bull fighting. Since they had the Aegean and much of the Mediterranean to themselves, Crete's great fleets brought in the wealth of a hundred shores.

With the rise of cultures on the mainland of Greece and in Phoenicia, Crete declined in importance, and by the time Athens was at its height of civilization the Minoans and their kingdom existed only in legend. It was not until 1893 that Sir Arthur Evans, a British scientist, unearthed the palace at Knossos and proved the existence of history's forgotten civilization.

For modern Greece, Crete produces great quantities of olive oil, fruit and vegetables. A local wine is made and the livestock and dairy industries are gaining importance. Deposits of iron and lignite are found on the island.

THE AEGEAN SEA

The Aegean, bounded on the north and west by mainland Greece, on the south by Crete and on the east by Turkey, was an early breeding ground for Western civilization. There is hardly a place on the sea where you would be out of sight of at least one Greek island—they cross the sea like stepping stones. These islands have traditionally been assigned to geographical and historical groupings: the Saronic Islands, the Cyclades, the Dodecanese, and the Northern Sporades.

The Aegean islands are for the most part dry and rocky. Farming is often difficult, but the stout Greeks have made several of the islands bloom. There is an infinite supply of fish in the waters, and livestock raising is gaining ground in places.

THE SARONIC ISLANDS

More than a score of islands, and countless islets, are within a few hours cruise from Athens, in the Saronic Gulf, a rectangular extension of the Aegean between the mainland and the Peloponnesus Peninsula. The gulf takes its name from the mythological King Saron who

The Palace of Minos at Knossos, Crete, as it looked when the Minoan civilization was at its zenith, between 1700 and 1500 B.C. The remains of this and other buildings on Crete indicates the Minoans had developed advanced engineering skills.

is said to have drowned there. These are among the smallest of the Greek islands, but because of their nearness to Athens, they are among the most popular with tourists.

AEGINA, a triangle of land less than 10 miles across, is the second largest, most popular, and easiest to reach of these islands. Lying exactly in the middle of the gulf, it is a rugged, hilly island with a population of more than 10,000. In 650 B.C. Aegina was the first European state to mint coins, and its currency was later accepted as the standard monetary unit throughout Greece. The main city and port of the island is also called Aegina. High above this city stands the awesome grey ruins of the Doric Temple of the goddess Athena Aphaia, built in the early 5th century B.C. Donkeys take visitors from the port up the steep trails to this remarkable site. Those residents not directly employed by the tourist trade usually grow pistachio nuts or make pottery.

HYDRA, one of the southernmost of the Saronic Islands, has long been a retreat for writers and artists. Its glistening, white villages against a background of blue-black sea have often been photographed and they provide many people's standard image of what a Greek island is like. Hydra is densely populated and dominated by a hearty breed of sponge fishermen who are dear to all Greeks for their part in the War for Independence from Turkey in 1821. It is the island's natural beauty rather than buildings which make it so popular, but there is a magnificent monastery atop the main hill.

POROS, perhaps the greenest of all islands in the Aegean, has a low, rolling landscape, totally blanketed with grass and trees. Barely a speck on most maps, it almost touches the Peloponnesus, between Aegina and Hydra. A monastery, ruins of a Greek temple, and attractive beaches make it an appealing place to visit.

SALAMIS is the largest and most populated island in the Saronic Gulf. The Greeks defeated a vastly superior Persian fleet here in 480 B.C., giving the island a significant, if short-lived, place in classical history. At the summit of the island stand the ruins of a great acropolis, or citadel.

Other important islands in this chain include METHANA, with its hot sulphur springs, ERMIONI, with excellent mosaics left by wealthy Roman vacationers, and SPETSAI, with prehistoric remains dating from 7500 B.C.

THE CYCLADES

The Cyclades are an archipelago containing many of the most familiar names among the Greek islands. They seem to be an extension of the mainland, spreading southeast from Athens. Homer called them the "highways of the fish."

The name Cyclades comes from the ancient Greeks, who thought their circular pattern, enclosing the sacred island of Delos, was not just a quirk of nature, but the work of the gods.

DELOS was said to have been chosen by

Like Mykonos, Ios, one of the smaller Cyclades, is made all the more interesting by thousands of old windmills that draw fresh water from deep wells.

Apollo, the sun god, as his landing place on earth, and, by Homer's time, it was the hub of all religious worship. So sacred was its soil that the pregnant and very old or infirm were banned so that the island would know neither birth nor death. Not only is Delos the heart of the Cyclades, it is also the midpoint of the Aegean. It is a barren rocky islet, and few of its supposedly subordinate sisters are so small. Almost the entire island is built up and islanders have guarded the ruins here almost constantly since ancient times. The ruins were left not only by the Greeks, but also by the Romans, who came in the next epoch, bringing Delos' greatest period of prosperity along with a decline in its religious importance.

Behind the old city of Delos, rises the mountain of Kynthos, the legendary resting place of Zeus. There are many old houses on the island that are open to visitors, but not many new ones. Few people live on Delos other than the guards who still watch over Apollo's temples after all these centuries.

Visitors may jet into Mykonos, but they usually end up on the slow, traditional donkey if they wish to see the most interesting parts of the island. Roads are paved with stones cut from the island, which, according to mythology, was itself a big stone thrown from Mt. Olympus.

MYKONOS is the most popular of the Cyclades among foreign visitors. The whitewashed town of the same name glows in the sunlight and can be seen far out at sea. Delos is within sight on the south. Mythology says that Mykonos was once a huge rock hurled by the Gods of Olympus at invading giants.

Buildings on Mykonos incorporate an interesting combination of cubes and arches, a style that has come to be known as Cycladic architecture. The streets are often too narrow for anything but pedestrian traffic, and are

The cubic houses and many little churches clustered around the waterfront of Mykonos are among the best examples of Cycladic architecture. A popular resort city, Mykonos is the home of many artists and a famous School of Fine Arts.

lined with picturesque arcades and miniature chapels.

Outside the town, windmills dot the terrain, along with an occasional fig grove or barley field. A wine produced on the island is prized by the local people, but has a limited market elsewhere. The highest spot on the island is Mount Ayios Ilias, a four-hour trek by mule from the town.

KEA, sometimes called Tzia, is one of the nearest islands to the Greek mainland. A small islet with a population of only 7,000, it claims three ancient cities—Korrisia, Ioulis, and Karthaia. Recent finds indicate that Kea's position in ancient times was of far greater importance than previously thought. On the hill that crowns the island is a medieval fortress built by the occupying Venetians. There are lush vineyards and orchards on Kea, and the sandy beaches are popular with swimmers.

PAROS (not to be confused with Poros, one of the Saronic Islands) is something of a holy spot for Greek Orthodox Christians. Here are the spectacular Church of Ekaptontapiliani and the Monastery of Loggovardi. Paros is said to have been the first of the Greek islands to accept Christianity. Quarries here supplied much of the marble for the glories of Athens and they can still be visited today.

The tiny town of Phia on the island of Santorini overlooks a bay that was once solid ground. About 4,000 years' ago, a volcano eruption and earthquake caused part of the island to fall below sea level. This same disaster may have brought an end to the Minoan civilization on Crete, some 75 miles to the south. Santorini is known for its steep hills and tiny villages perched on treacherous peaks. A volcano crater which still smokes adds to Santorini's unique charm.

NAXOS is the largest, and some say most beautiful, of all the Cyclades. The town of the same name is medieval, with castles left by ruling Venetian dukes. Open-air markets here have a Middle Eastern atmosphere. Lemons provide Naxos' main source of income.

SANTORINI, the most extraordinary of the Cyclades, is the collapsed cone of an ancient volcano and the circular port was once the volcano's core, thought by some islanders to be bottomless. The town here is called Thera. Some scientists have proposed the idea that the "lost continent of Atlantis" was Santorini. The collapse of the volcano, which greatly reduced the size of the island, might have inspired the legend of Atlantis sinking into the sea.

Other important Cyclades include the picturesque MILOS, which gave the Louvre Museum in Paris its most famous piece of statuary, the Venus de Milo; KIMOLOS, in the western Cyclades, with ruins from every civilization from Mycenean to Byzantine; ANDROS, the northernmost island, famous for its lemons, almonds, and olives; and IOS, the supposed site of Homer's tomb.

THE DODECANESE (Southern Sporades)

Dodecanese means "twelve islands"—but counting islands must have been difficult in the ancient days when the name was first given, for there are actually 20 islands and islets in the

This old etching shows how the great Colossus of Rhodes stood at the narrow port entrance. About 230 B.C., an earthquake sent the 100-ft.-high statue tumbling into the water. Later it was removed and taken to Turkey, where it was probably melted down and the bronze sold.

chain. They stretch along the Turkish coast far to the east of the Greek mainland and are among the more remote Aegean islands.

The acropolis at Lindos, which includes the temple of Athena Lindia, rests on one of Rhodes' seaside hills. Just beyond these columns, the land gives way to a sheer 400-foot drop to the sea, making the location ideal for defence. Lindos was one of the three original cities on Rhodes.

The modern city of Rhodes stands on the northwestern coast of the island of the same name, outside the walls of the old town built by the Knights of St. John of Jerusalem.

This island group has lived through the glories of Greek and Roman civilizations and the indignities of repeated occupations. From 1911 to 1945, they were held by the Italians, a grievance that the 123,000 people living on the islands have never forgiven. An earlier defeat came in 1522 when the Turkish Sultan, Suleiman I, won the islands for the Ottoman Empire. The main sources of income for the people of the Dodecanese are livestock, sponge fishing, and some agriculture.

RHODES, one of the world's best known islands, lies only 9 miles from Turkey. More than 50 miles long, it is the largest of the Dodecanese group and the most eastern of any major Greek island.

In prehistoric times the island was settled by the Cretans. About 1100 B.C., Dorians, a Greek tribe from the mainland, built three cities, Ialysos, Camiros, and Lindos, all of which became major ports. In 408 B.C. these cities' leaders united to build a new capital on the northern point of the island. The result was the city of Rhodes, often called the most beautiful city in the world by Greek and Roman historians. Its port was guarded by the 105-foot-high bronze Colossus of Rhodes, one of the "Seven Wonders of the Ancient World." It took 12 years to build this massive figure, which stood only 55 years before an earthquake sent it crashing into the water. The island of Rhodes, along with its sister islands, later became part of the Roman, then Byzantine, Empires. In A.D. 1309, it was sold to the Knights of St. John of Jerusalem, the militant crusaders who ruled as absolutely here as they later did on Malta.

The modern city of Rhodes is the commercial, transportation, and governmental headquarters for all the Dodecanese. The old

13

Skopelos has a shoreline indented with hundreds of small bays and coves, most of which have at least one bright whitewashed chapel shining against the green landscape. Above is shown the Paraghia Chapel, whose grey shale roof is typical of most buildings on the island.

cities of the island, especially Lindos, have hundreds of archeological sites of interest.

A unique area on the island is Petaloudes, the Valley of the Butterflies, where the sweet scent of flowers attracts thousands of bright butterflies. They often fly in legions so thick that they darken the landscape.

PATMOS, the most northern of the group, is almost as sacred today as Delos was in antiquity. Here John the Apostle wrote the Book of Revelations. The site of the writing is marked by a basilica. There are many small bays around the edge of this small island, and two popular beaches at Griko and Kikoffi.

LEROS, just to the south of Patmos, is rocky and arid with steep cliffs on the coast and irrigated vineyards inland. The bays here contain many tiny islets of their own.

KOS, the birthplace of Hipprocates, who gave medical doctors their code, is rather large, very fertile and almost covered with ancient and medieval relics. There are mineral springs in the inland mountains and several well developed resorts along the coast.

Other members of the Dodecanese, all small, rocky islands of brown stone, with white villages overlooking blue sea, are KALYMNOS, NISIROS, TILOS, CHALKI, and SYMI.

THE NORTHERN SPORADES

The Sporades have been grouped in various ways by different geographers. They are usually divided into two groups—the Southern Sporades, which include the Dodecanese and some outer islands, and the Northern Sporades, which include most, if not all, the islands in the Northern Aegean. Euboea, second only to Crete in size among the Greek islands, is sometimes excluded from this group, and Lemnos, Hilos, Lesbos and others far to the north or west of the Aegean, are often excluded.

EUBOEA. But for a few feet of water, imposing Euboea would be a great peninsula. About 90 miles long and from 4 to 30 miles wide, the island runs north to south, flanking over half of the mainland's Aegean coast. At two places it nearly touches the mother country and nowhere does it disappear out of sight. Here are high wooded mountains rising to a height of 7,000 feet, with splendid views of the Aegean. The main town, once capital of an ancient city-state, is Chalcis. The city faces the mainland across the Euripus, a narrow strip of water spanned by a drawbridge. The channel is more famous than the city because of the curious fact that the current reverses its direction on the average of 14 times every day. Scientists are at a loss to explain this fact and the philosopher Aristotle is said to have drowned himself because he could not solve the mystery.

Most other towns on the island are fishing villages, which also face the mainland. The eastern coast is rugged and treacherous and has been a hazard to shipping in the Aegean throughout history. The population of Euboea is about 166,000, the majority of whom are employed in cattle raising. Tourism has not arrived on Euboea on a scale large enough to alter the natural beauty and atmosphere. The island passed through the hands of each civilization touching the Aegean. During the Venetian period it was called Negroponte, which still survives as a second name.

SKIATHOS lies a few miles north of Euboea, also quite near the mainland. Its capital is built on the sides of two low hills, with red-roofed houses clinging to narrow, steep streets. Its former capital, Kastro, is almost inaccessible because of a deep ravine, but the grim old walls can be seen from one of the island's roads. Pines cover much of the land, and gracefully curving beaches completely surround the island. Fishing is the only industry other than tourism.

SKOPELOS. At one time nearly every island in the Aegean had a capital city named Chora—Skopelos is one that still does. There are 360 churches on the island, many only one-room chapels. The architecture here is unique, most buildings being three-storeyed with distinctive grey tile roofs. On a dozen inland hills are found forgotten convents that date from the 17th century. Several of these contain valuable paintings, carvings, frescoes, and icons.

ALONNISOS. The ruins of cities which sank into the sea are visible through the clear waters surrounding Alonnisos. This little island also has a town named Chora atop its central hill. Lobsters are caught here for export, as are octopuses.

LEMNOS, east of Euboea, claims to have been the home of the god Hephaestos. Its beautiful capital, Myrina (now called Kastro), was built on the ruins of an even more ancient city. A sweet red wine is an important export.

CHIOS is a big, very fertile island south of Lemnos, famous for a gum exuded from the mastic tree. This is used in making liqueurs and as a base for chewing gum. The port city of Chios lies on the eastern coast, facing Turkey, the country that ruled the island for centuries. Two historical sites on the island include a stone said to have been used by Homer as a chair while teaching his students and an 8th-century church containing several priceless Christian relics.

LESBOS, a large island close to the Turkish shore, is famous as the home of the woman poet, Sappho. A rugged, but fertile land, Lesbos produces wine, olive oil and grain. Mytilene, the chief town, was an important city of the Byzantine Empire, before coming under Venetian, Genoese, and Turkish rule. It was not returned to Greece until 1912. Other important

In the bay of Corfu, the historic old island monastery of Pontikonnissi, one of the chief landmarks of the Ionian Islands, greets visitors arriving by sea. Beyond the monastery is the cypress-covered Mouse Island, a retreat for poets and site of an old bell tower which tolls each evening at sunset.

islands in the area are IKARIA, SAMOS, THASOS, and SKYROS.

THE IONIAN SEA

To the west of mainland Greece lies the blue Ionian Sea, a wedge of the Mediterranean bordered on the west by Sicily and the sole of Italy's boot. Seven large islands and numerous islets line Greece's Ionian shore, Corfu being by far the most famous of these.

The Ionian Islands, scattered in a gently curving line for 325 miles, have a total area of 1,117 square miles. They are more rugged and mountainous than the islands in the Aegean, and are also greener. Wheat, olives and grapes grow on the narrow coastal plains. The Venetians had a long, successful rule here from 1386 to 1797, and their influence can be felt everywhere. There was a brief period when the Ionian Islands became a free and independent state. From 1815 to 1864, England ruled the

The island of Sifnos has known days of great prosperity, when its mines were rich in gold and silver. Today, however, its 2,000 inhabitants are mostly fishermen, farmers, or potters. On the hill beyond, farmers have built terraces to make more flat land for growing their grapes.

islands. Then, in 1864 they were returned to their mother country, Greece. Today they form a province of Greece.

CORFU. The name was a gift of the island's medieval Italian rulers, but island history predates its name by 1,000 years. In Homer's *Odyssey* it was called Scheria, the home of the Phaeacians. The Greek city-state of Corinth colonized the island in the 8th century B.C., but the islanders later entered into an alliance with Athens. Next Rome, then Constantinople, took charge. With the help of the Venetians, Corfu was able to withstand repeated attacks of the Turks, so it is one of the few islands in the Mediterranean with no period of Islamic rule.

Corfu is the northernmost island in the Ionian chain and lies very near the coast of Albania. Its highest point is Mt. Pantokrator (3,000 ft), but most of the island is low and fertile. Livestock and fishing share with olives important places in the local economy. Massive groves of olive trees give the entire island a silvery-green cast. Some citrus fruit and grapes are also grown for export.

The town of Corfu is the provincial capital of the Ionian Islands. In contrast to the white-washed villages of the Aegean, it is a veritable rainbow—pink and yellow houses against dark green hills surrounded by a deep blue water. The old fortress and magnificent town hall were built by the Venetians, as were many of the hundreds of charming tiny streets running up the hillsides. There is a 16th-century cathedral and a shrine to the island's patron saint, St. Spyridon, whose silver casket is paraded through the town four times a year.

At Mon Repos, near the city, are remains of a temple built in the 6th century B.C. to Asclepius, the god of healing. Many other ancient ruins, including entire cities, are being unearthed on the island, and within a few years Corfu could be one of the most important archaeological sites in Greece. PAXI, a tiny island covered with semi-tropical vegetation, lies just south of Corfu.

LEUCADIA was famous in antiquity as the scene of Octavian's naval victory over Mark Antony and Cleopatra, although that battle takes its name from Actium, a town on the mainland. Today it is best known as the site of a world famous festival of art and literature.

CEPHALLONIA is the largest and most centrally positioned of the Ionian Islands, but not so popular with tourists and scholars as Ithaca, nearby.

ITHACA, a mere sliver between Cephallonia and the mainland, is mentioned often in classical literature. One of the ruins here is claimed to be the castle of Odysseus, hero of Homer's *Odyssey,* though most scholars contend he never lived and that Leucadia better fits the descriptions of the island called Ithaca by Homer, than does Ithaca itself.

ZANTE is a garden of bright flowers, rare paintings, and Christian relics. Its pitch mines were mentioned in writings dating back to the 5th century B.C.

KYTHERA, far to the south, has mostly shared the history and fate of the Peloponnesus, to which it belongs geographically. Called Cythera in English, the island is associated with the birth of the goddess Aphrodite, and to students of painting is famous as the subject of Watteau's work, "The Embarkation for Cythera," now in The Louvre in Paris.

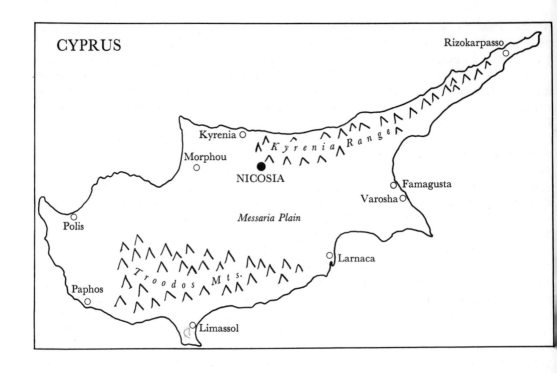

CYPRUS

CYPRUS

ACCORDING TO legend, Cyprus was the home of Aphrodite—the Greek goddess of love. Majestic mountains, golden beaches, and picturesque walled cities make this, "The Island of Love," seem all the more romantic today. Centuries of political struggle have left the little republic with an image more often associated with war than with romance, however.

GEOGRAPHY

Cyprus lies near the eastern extremity of the Mediterranean, just 40 miles south of Turkey and 60 miles west of Syria. It is about the same size as Corsica (3,500 square miles). The Troodos or Olympus Mountains (not the same as the famous Mt. Olympus in Greece) in the

southwest and the Kyrenia range along the northern coast are linked by the beautiful Messaria plain. Much of the land was once covered by forests which have now given way to cities, orchards, and farm lands. The 486-mile coastline is highly indented and rocky, but here and there sandy beaches are found.

The Messaria presents a dazzling pattern of small farms and vineyards along the banks of the two main streams, the Psedias and the Yalias, both of which flow east through the plain. Herds of goats and sheep often block the narrow roads of this region.

Nicosia, with a population of more than 100,000, is the largest city and has been the island's capital since A.D. 1200. In the heart

Cube-shaped houses on curving streets make an interesting pattern at Lefka, a farming town on the Messaria plain. Beyond the town lie vegetable fields and in the foreground is the church, built in typical Cypriot Orthodox style.

Sheep farming has been a way of life on Cyprus for many years, even though much of the land is poorly watered and pastures periodically dry up. This Cypriot shepherd will soon benefit from an irrigation system stemming from the Ovgos Dam—a spillway for the system can be seen at the bottom of the hill. The new dam is a project of the Cyprus Water Development Department, a government agency that is making great progress in easing Cyprus' ancient thirst.

The skyline of Kyrenia includes many minarets, or prayer towers, of Moslem mosques. This is a predominantly Turkish Cypriot city, and the scene of some of the country's most violent anti-Greek uprisings.

of the Messaria, Nicosia has many modern buildings that contrast starkly with the well preserved old section. Flowers now grow in the moat that once protected the population from never-ending invasions.

Other important cities include Famagusta, once the most important port in the eastern Mediterranean and thought to be the scene of the main action in Shakespeare's *Othello,* and Kyrenia, a seaport on the northern shore whose ancient buildings face the Turkish coast a short distance away. Larnaca, a southern seaport, is a holy city to the Islamic world.

Since ancient days, Cyprus' location along the Mediterranean sea routes has made the island of vital importance to anyone who wished to control the sea. While an asset to the economy, this strategic location has brought war to Cyprus and kept its people under foreign rule throughout most of its history.

HISTORY

The oldest evidence of civilization on Cyprus is found in ancient pottery and stone carvings that date from about 3700 B.C. The Egyptians, who were first to mention Cyprus in records, called the island Asi, and they dominated early trade with the Cypriots. Recently evidence was uncovered that suggests the Cypriots had

commercial contact with the Greek islands as early as 1500 B.C.

In 709 B.C. Cyprus submitted to the rule of the Assyrians. Control of the island later shifted to the Persians and Greeks. Augustus Caesar claimed Cyprus as his personal property, and the Romans ruled it for 450 years. When the Empire was divided, Cyprus went to Byzantium in the East. It was taken by Richard the Lionhearted for England in the 12th century. Richard turned his conquest over to the former Crusader King of Jerusalem, Guy of Lusignan, whose family ruled the island for the next 300 years. As the Lusignan dynasty declined, Cyprus fell prey to foreign invaders. For years Venice and the Turks fought for Cyprus with neither claiming absolute control for over a century. In 1489, Venice took over the kingdom, and then, in 1571, the Turks finally won control of the entire island. The British occupied Cyprus in 1878, annexed it in 1914, made it a Crown Colony in 1925, and remained on the island until independence in 1960.

Everybody who claimed Cyprus left landmarks that give the landscape a unique quality. Ancient pagan ruins, Greek and Roman cathedrals, mosques, medieval castles, and British colonial buildings stand side by side here.

THE PEOPLE

Cyprus' greatest challenge today is not to resist invading powers but to maintain the unity of its own population. Of the 600,000 people living on the island about 80 per cent are of Greek origin while 18 per cent are of Turkish ancestry. Each group has inherited different languages, religions and traditions. Most of the

Archbishop Makarios is both president of the republic of Cyprus and head of the Cypriot Orthodox Church. The Archbishop has pursued a policy of reconciliation, attaching himself to neither Greece nor Turkey, and has worked to guarantee full rights of citizenship to Turks. The country's House of Representatives has 35 seats for Greek Cypriots and 15 for Turkish Cypriots, though the Turks have boycotted the government entirely since 1964.

The medieval castle of St. Hilarion is a relic of the days when Venice and Turkey did battle for control of Cyprus. Situated in the Kyrenia mountain range on the northern coast of Cyprus, the castle can be seen from the mainland of Turkey on a clear day—which is not unusual in the eastern Mediterranean.

21

The life of the Cypriot peasant goes on much as it has for centuries. These women take their flocks of sheep out each day to graze on the hills near Kato Pyrgos. It is not uncommon for farmers and shepherds to live in small towns several miles from their fields or pastures.

Greek Cypriots stand guard on a hill near positions occupied by Turkish Cypriots.

Following repeated outbreaks of violence on Cyprus in 1964, the conflict threatened to spread to Turkey and Greece. The United Nations, in one of its most decisive moves, sent an international peace-keeping force comprising soldiers and civilian police from several neutral countries. About 4,000 foreign troops still remain on Cyprus, including these members of the Canadian contingent stationed at Kyrenia.

Greeks are members of the Orthodox Church of Cyprus (Christian), while the Turkish Cypriots are Islamic and represent the only significant influence the once powerful Moslems still exert in the Mediterranean islands. Though Greek and Turkish are the native languages, English is the *lingua franca*.

Many Greek Cypriots have long pressed for *enosis*, or unity with Greece, while the Turkish population has demanded the partition of the island between Turkey and Greece. The dispute erupted into serious violence in the 1950's and was finally resolved in 1960 when Cyprus became an independent republic with no political ties to Greece or Turkey. Archbishop Makarios III, who first proposed the settlement, is both Primate of the Cypriot Church and President of the Republic.

While the Archbishop's government has

These school children from the Turkish Cypriot community in Geunyeli are among nearly 150,000 enrolled in the young country's 848 primary and secondary schools. The brightest hope of unity among all Cypriots seems to rest with this young generation.

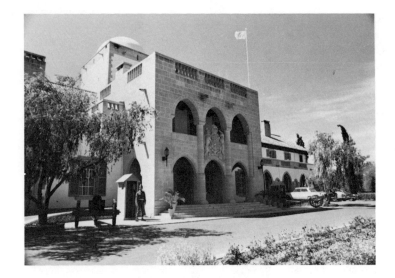

The official residence of Archbishop Makarios is in Nicosia.

survived for over a decade, peace has never come. The Turkish population resent its poorer standing and accepts financial aid from the government of Turkey. Independence has not silenced the cry for *enosis*—in 1970, militant Greek Cypriots attempted to assassinate Archbishop Makarios, nearly causing a crisis in the government.

THE ECONOMY

Like most of its sister islands, Cyprus depends upon agriculture for its income. Throughout the plains, Greek and Turkish Cypriots cultivate the soil side by side. Many own their own land, and farming methods have been modernized since independence. Major export

Dissatisfied Turkish Cypriots demonstrate against the United Nations forces. Though they are guaranteed full freedom by the government, the standard of living among Turks is much lower than that of their Greek compatriots.

Citrus fruits grow large and sweet in the warm Cyprus sunshine, and they are an important part of the island's export trade. Britain, West Germany, Italy, France, Russia and Greece are among the island's main customers. These Cypriot farmers are picking and crating oranges for export.

→

← These Cypriot farmers are learning new techniques with the help of the United Nations' Food and Agriculture Organization.

Fruits are also popular among the Cypriots. Here a vendor sells lemons and newspapers in a Nicosia market.

To help increase the supply of manpower for its new industries, Cyprus established a Productivity Centre several years ago. This agency provides information, advice, and training for both public and private industry, aided by the United Nations.

Open-pit mines like this one at Xeros are used to extract the rich deposits of copper on the island of Cyprus. Cyprus has been a major source of copper since ancient times—in fact, the island takes its name from "kypros," the old Greek word for copper.

At the furniture factory pictured here, productivity experts have helped to streamline and standardize production methods.

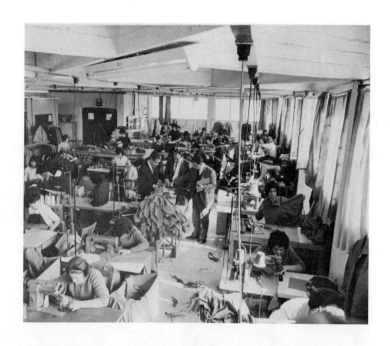

The textile industry is growing in importance in Cyprus as irrigation makes more of the land arable. This clothing factory in Nicosia, the largest of many modern mills, uses locally grown cotton to produce textiles for export.

crops include potatoes, citrus fruits, raisins, and carobs, the last named being the seed pods of a tree of the pea family, used for fodder and human food. Grapes, wheat, barley, and other vegetables are grown as food crops. Cotton is becoming more and more important as the textile industry grows.

Fishing is not as important to the economy of Cyprus as the industry is on many other Mediterranean islands. In fact, fish has to be imported just to meet the local demand. Sponges of high quality are taken from the shallow sea floor near the coast.

Snow-covered Mt. Etna provides an unexpected background for warm beaches near the hilltop city of Taormina. This photograph, taken from the town, shows the honeysuckle vines and fruit trees in bloom that give the eastern coast of Sicily a unique, subtle quality.

SICILY, SARDINIA, AND OTHER ITALIAN ISLANDS

OF THE VARIOUS islands in the Mediterranean, Italy owns the two giants, Sicily and Sardinia, both over 9,000 square miles in area. Other Italian islands are those of the Tuscan archipelago, the Aeolian and Pelagian groups, and the isolated islands of Ustica and Pantelleria.

SICILY

If the Italian Peninsula is thought of as a giant boot, Sicily is the rough stone it kicks. This grandfather of the islands lies just off Italy's "toe," separated from the mainland by the narrowest of channels. The ancient Greeks thought the island's triangular shape was its most characteristic element, so they dubbed it Trinacria, "the three cornered." The visitor to Sicily, however, will probably remember the island most for its spicy cuisine, breath-taking panoramas, and warm, hospitable people.

TERRAIN

Over 180 miles from east to west and almost as wide at some points, Sicily's vast moun-

Where lava from Mt. Etna has reached the Ionian Sea, small promontories, scenic grottos, and surrealistic rock formations have displaced the more common sandy beaches along the Sicilian coastline. This area, with its shallow, protected coves, is called the "Riviera dei Ciclopi," or "shore of the Cyclops."

tainous landscape is home to nearly 5,000,000 people, more than any of its sisters in the Mediterranean. Just 2½ miles of shallow sea, the Strait of Messina, separates the beach at Ganzirri from the mainland. Tiny Malta and the coast of Africa lie to the south, with Sardinia over the northwest horizon. The Mediterranean waters to the north of Sicily are called the Tyrrhenian Sea, and to the east, the Ionian Sea.

Towering mountains, an extension of the Apennines that form the spine of the Italian Peninsula, and Mount Etna, an active volcano, give Sicily a rugged, irregular face with thousands of green valleys.

Nature and man have clashed violently several times in Sicily's history. Mt. Etna's eruptions have occasionally been disastrous, most recently in the spring of 1971. Earthquakes have periodically levelled all but the soundest structures.

The rocky slopes of the mountains are

In Palermo the sweet scent of oranges, fresh from the inland orchards, drifts through the streets near the markets. These men are wrapping and boxing the fruit for shipment to Rome.

Ragusa is a picturesque town situated on two levels; the lower town was mostly destroyed by an earthquake in 1693. Later, the people moved to the plateau above and rebuilt their town. Though this is an old site, most of the architecture, even in the old town, dates only from the 18th century.

blanketed with wild flowers in warm months and with snow on the upper levels during the winter.

Nine of Italy's 91 provinces are found on Sicily. All the chief towns are ports, but they are linked by an impressive network of roads and, unusual for Mediterranean islands, by railways that cross the hinterland.

CITIES

Palermo, with a population of 600,000, is Sicily's largest and busiest city, and serves as regional capital. An important port city since ancient times, it lies about 140 miles west of the Strait of Messina, cradled in a valley called the "Golden Shell." Architecture here is as diverse in style as the civilizations that have occupied the land. Several impressively ornamented churches show a unique blending of Eastern and Western tastes.

Catania, second in size of Sicily's cities, lies on the eastern coast at the base of Mt. Etna. Ancient Greek and Roman ruins sit side by side, including an aqueduct, a theatre, and a vast Roman bath.

Messina, though hampered through history by earthquakes, has survived it all to become one of Sicily's most modern cities. Lying near the Strait that bears its name, Messina is the gateway to Sicily from the mainland.

There may be no city in the world that occupies a more unlikely position than Taormina, which is literally plastered along towering narrow terraces overlooking the eastern coast. The view from here is spectacular and tiny winding streets offer unique settings for the creative photographer.

Syracuse, near the southeast corner, dates from the arrival of the first Carthaginians in

Completely surrounded by mountains and the Mediterranean, Palermo offers a breathtaking panorama from any vantage point. The plain on which the city lies is often called Sicily's "Golden Shell."

734 B.C. and was already an antique city when the Roman Empire was born.

Countless other villages around the island were once important and their stately ruins offer an interesting contrast to the simple life of the modern villagers.

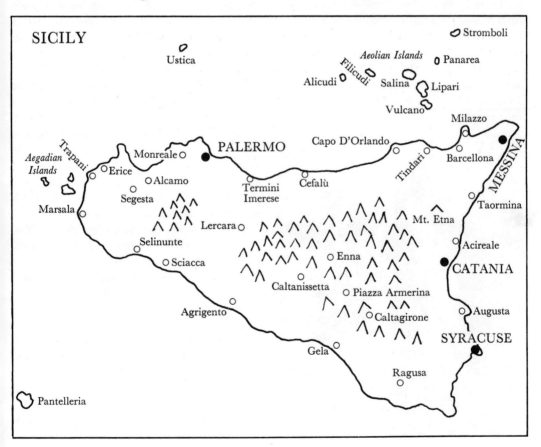

SICILY

Stromboli

Ustica

Aeolian Islands

Panarea

Alicudi Filicudi Salina Lipari

Vulcano

Milazzo

Capo D'Orlando

MESSINA

Aegadian Islands

Trapani

Monreale

PALERMO

Barcellona

Tindari

Erice

Alcamo

Cefalù

Termini Imerese

Taormina

Segesta

Marsala

Mt. Etna

Lercara

Acireale

Selinunte

Enna

CATANIA

Sciacca

Caltanissetta

Piazza Armerina

Augusta

Agrigento

Caltagirone

SYRACUSE

Gela

Ragusa

Pantelleria

Ruins of the Roman Temple of Juno, mythological queen of the heavens, sit atop a hill near Agrigento. One of hundreds found along the southern coast of Sicily, the temple was built about 500 B.C. Nearby are equally impressive temples of Jupiter, Vulcan, and Hercules.

HISTORY

Sicily had not yet been separated from the mainland of Europe when the first men arrived in the early Stone Age. Descendants of these early Sicilians gradually developed a civilization that easily adapted to the Phoenician, Greek and finally Roman cultures that were brought to its shores. During the Golden Age of Greece, the island's hills were adorned with splendid temples. No god of any stature was without a Sicilian temple.

The Punic Wars, which led to Roman supremacy in the Mediterranean, were fought over ownership of Sicily. When Rome reached its peak, so did Sicily, with marble cities rivalling any in the Empire. With the rise of Christianity, the pagan temples became churches and were later adorned with spectacular mosaics during the years of Byzantine rule.

In the 9th century, the Arabs took Sicily from Byzantium. Their rule, though short, was a time of great advance in commerce and in science. In 1130, Roger II, a Norman, became the first king of Sicily and the island became the Normans' prized possession. Later, the Spanish monarchs, with the aid of the Papacy, took over, and ruled the Sicilians harshly. In 1816, Sicily and Naples, united at different times since the Middle Ages as the Kingdom of the Two Sicilies, once more became the Kingdom of the Two Sicilies. The union lasted until 1860 when Giuseppe Garibaldi conquered the island and paved the way for union with Italy.

Like all Italians, Sicilians love art and tradition. The two are pleasingly combined in the native costumes of each province, as with these happy people from Messina. Today, the costumes are mostly reserved for festivals, but in remote villages they have never given way to modern styles.

The most remarkable structure left by the benevolent Norman rulers of Sicily is the sprawling Cathedral of Monreale, 4 miles outside Palermo. Through earthquakes and political turmoil, the cathedral has retained its towering beauty.

One of the most impressive churches on an island famous for its religious architecture is the Church of St. George in Modica, a town near the southern coast of Sicily. Built during the 18th century, this baroque landmark stands at the head of an imposing flight of steps.

Palermo has about 600,000 people and on any of several major religious holidays just about every one of them will be in the city's narrow streets to join in the celebration. Sicilians are very much Italian, but they also have their own traditions in literature, music and dress. The hand-embroidered dresses worn by these Sicilian women follow a style that has its origins in the Middle Ages. They show the Spanish influence Sicily felt during the latter part of those dark years.

The majestic campanile of Messina's cathedral rises above vaults that contain precious Christian relics. The Norman influence here is pronounced. In the foreground is the Fountain of Orion, the work of the Florentine sculptor, Montorsoli.

descendants of every civilization that came to the island's shores and, at the same time, very much Italian.

Sicilians especially like the arts, and people from all walks of life pour into the cities during the spring for numerous operas and symphonies. They are also devoutly religious, setting aside several weeks each year for sacred celebrations.

THE ECONOMY

Agriculture is the main source of income for Sicilians, and more of the island is being planted as irrigation projects are completed. Because the weather is always warm, there are as many as three harvests each year. Citrus fruits are among the products Sicily is best known for, and the sweet smell of lemons and oranges is in the air for much of the year. Grapes and wines are also exported, along with processed fish. Sulphur and petroleum are gaining importance on the island, as is the related chemical industry. And, of course, tourists provide much of the island's income.

THE PEOPLE

Culture has always flourished among Sicilians, even during the Dark Ages. Today these proud people guard their heritage, a combination of many elements, against the threat of a world that changes too fast for them. They are the

Stromboli is an active volcano that rumbles on the inside but remains one of the world's most peaceful islands on the exterior. Great flocks of birds make their homes on the 3,000 foot cone.

There are at least as many fishing boats as there are people in the village of Mondello, whose crystalline bay is protected from the trade winds and storms by Mt. Pelegrino. Although a popular beach resort has grown up at the edge of the bay, the village has been almost unaffected by the development, retaining its unique charm.

ISLES OFF SICILY

Sicily is bordered on three sides by archipelagos and solitary islets, all closely associated, geologically and historically, with the main island.

Most important of these are the AEOLIAN ISLANDS, sometimes called the LIPARIS. They are unusually interesting for their rocky lagoons, active volcanoes, and important position in Roman mythology. All the islands lie in the Tyrrhenian Sea, northwest of Messina.

VULCANO is noteworthy for its sometimes violent volcano, where Vulcan, god of fire, was said to live. Aeolus, after whom the chain was named, was another legendary being who called Vulcano home. In Greek mythology he was keeper of the winds, which he kept confined in a cave.

PANAREA, LIPARI, SALINA and STROMBOLI are lesser members of the group, as is FILICUDI, which enjoyed temporary fame in 1971. The Italian Government exiled 15 famous gangsters to Filicudi. The islanders protested and, when the government refused to remove the newcomers, all 270 Filicudians left their island home, threatening not to return until the undesirables left.

Off the western corner of Sicily lie the AEGADIAN ISLANDS, including FAVIGNANA, LEVANZO, and MARETTIMO, which are known for extensive tuna fishing and canning.

The Lake of Flumendosa near Nuoro is one of the highest inland bodies of water on Sardinia. Fed by mountain streams, the lake is tapped for irrigation and watering herds of sheep.

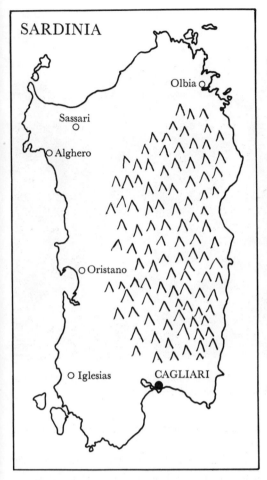

SARDINIA

Olbia
Sassari
Alghero
Oristano
Iglesias
CAGLIARI

The coast of Sardinia is almost everywhere rugged—at places sheer rock walls rise from the sea to several hundred feet. There are also a few sandy beaches, especially in the north. Evergreens and subtropical vegetation are found on the coastal plains and shrubs dominate the higher slopes.

USTICA, an island that sits alone in the Tyrrhenian Sea north of Palermo, offers one of the world's most scenic areas for skin diving.

PANTELLERIA and the PELAGIAN ISLANDS lie to the south of Sicily—the latter even south of Malta. They are visited less frequently than most of the small islands, but Pantelleria is famous for its raisins and prehistoric remains.

SARDINIA

Unlike Sicily with its vast population, Sardinia's countryside is austere and silent. On the second largest Mediterranean island (9,302 square miles) there is a population of under 1,500,000, most of whom are clustered in and near seacoast villages. Here you can walk for

miles along rocky paths and never meet another person. But you cannot walk very far without encountering an historical landmark, and nowhere is the scenery less than spectacular. Sardinia is one of the truly unspoiled treasures of the Mediterranean.

TERRAIN

Hundreds of millions of years ago there was a land area we now call Tyrrhenis, where the Tyrrhenian Sea is today. As Europe emerged from the seas, Tyrrhenis disappeared, leaving only the highest peaks exposed—that is Sardinia.

The island is still a rugged land, with mountains that rise to over 6,000 feet above sea level. Along the eastern shore, towering granite and gneiss cliffs look toward Rome, 165 miles away.

A bluff on the coast of Sardinia near Cagliari reveals the layers of sediments that originally formed the land mass of Tyrrhenis—most of which now lies beneath the Mediterranean.

Corsica, once joined to Sardinia, is now separated by the narrow Strait of Bonifacio on the north. Tunisia in North Africa is 130 miles to the south.

Sardinia is roughly rectangular in shape, 150 miles north to south, and 80 miles east to west, though the ancient Greeks, with their vivid imaginations, insisted it looked just like a shoe and called it Ichnousa, which means footprint.

Due to the stony soil and prevailing dry winds which blow across from the Sahara, Sardinia is less green than many islands in the Mediterranean, but with a wild beauty all its own.

Like Sicily, Sardinia is a self-governing Region of Italy. It is divided into three provinces: Cagliari, Nuoro, and Sassari.

The city of Cagliari, with a population of 200,000, is the island's capital and dominates a gulf on the southern shore. The history of this city is older than that of Rome, and many examples of ancient architecture are found throughout the surrounding province. Cagliari provides most of Italy's marine salt—in the countryside outside the city, man-made lagoons are used to evaporate sea water and extract the salt.

The church of San Saturno is from the early Christian era and is built on the original site of the martyrdom of the Saint. Another landmark of Cagliari is the university, founded in 1620 by Philip III of Spain.

Olbia, a northern city of about 13,000, is the closest port to Rome, and therefore many visitors land there. Sassari, also on the northern coast, is an important market town for agricultural products. Some of the finest medieval architecture in Sardinia is found there.

Nuoro is the only sizeable town not situated on the coast. It sits in the hills 40 miles from the sea and boasts the most healthful climate on the island.

HISTORY

A race of Sardinians once lived, and then vanished long before written history gave the island its most interesting relics. These Proto-Sardinians, as they are called, landed on Sardinia about 3,000 B.C. and began building

At the small village of Silanus, in central Sardinia, two of the island's best preserved historical sites stand within a few yards of each other. The "nuraghe" here is one of the most perfect still standing. Nearby is a church built between 800 and 900 A.D.

monuments called *nuraghi*. The *nuraghe* (*nuraghi* is the plural in Italian) is a hollow structure of rough, unhewn stones piled layer on layer to form a round tower with inward-

A crumbling old tower is all that remains of fortifications built on the western coast of Sardinia near the town of Alghero by the House of Aragon. Cape Caccia, almost 600 feet high, rises in the background. Along the rocky shoreline of this bay is Neptune's Grotto, reached via a steep "goat's stairway." People in this area still speak the old Spanish tongue of the Aragon landlords.

Sardinians have the raw materials, talent, and patience to produce fine handicrafts. This man of Sassari is making a fishing net from local thread.

This young man is wearing Sardinian folk costume for a festival.

sloping walls. Of the 8,000 of these towers that once stood, about 6,500 still exist in varying degrees of disrepair. Some are only a few feet off the ground, while others reach 60 feet or more. The purpose of the *nuraghi* is still in dispute—no buildings quite like them exist anywhere else on earth.

Given its central position in the Mediterranean Sea it was of course unlikely that Sardinia could remain outside the wake of advancing civilizations. Phoenicians, Carthaginians, Greeks, Romans, Vandals, the Popes, and the Spanish Crown all had their periods of prominence on the island. Some structures or ruins remain as the legacy of each, though none of the historical buildings match the *nuraghi* for endurance.

THE PEOPLE

Evidence indicates that religion has played a very large part in the lives of Sardinians since 2000 B.C. Prehistoric cult sites, ruins of classical

The Sagra of Sant'Efisio, celebrated each year on the first of May in Cagliari, is a time when people from all parts of Sardinia don their ancient costumes and come to the capital for the island's biggest holiday. It commemorates the Roman general who suffered martyrdom on the island under Roman Emperor Diocletian, and dates from 1656, when the citizens of Cagliari made a vow to venerate the Saint for having saved them from the black death.

A ship lies at anchor off Marina Grande, below the rugged cliffs of Capri.

pagan temples, and churches dating from the very early days of Christianity all testify to this. And so it is today. In Sardinia, life is closely tied to the Roman Catholic Church, and Sardinians reserve their most vivid celebrations and costumes for the numerous religious holidays. The medieval practice of offering candles to the Virgin has survived in the Sassari Festival of the Candlesticks, a celebration which shows, in its liveliness and warmth, a Spanish influence. Every locality in Sardinia has its patron saint, and each saint has a number of celebrations throughout the year.

No single occupation dominates the economy and the lives of the Sardinians, though agriculture and stock raising are more important than industry. Citrus fruits, olives, grains, potatoes, sugar beets, tobacco, and linen are the most important products of agriculture. What is not consumed on Sardinia is shipped to the mainland. Much land is given to pastures and meadows for the raising of cattle, sheep, pigs, goats, and horses. Many Sardinians make their

living by fishing, gathering cork, or mining salt, silver, or lead ore.

In the villages, home industries include the manufacture of fine carpets, gaily decorated baskets, ceramics and jewelry.

ISLES OFF SARDINIA

There are several small islands that lie just off the coast of Sardinia. SAN PIETRO and SANT' ANTIOCO, near the Gulf of Cagliari, are especially interesting for their volcanic rock formations, marine coves, and Punic temple ruins.

The castle of Ischia was built by the island's Aragonese rulers. It stands on a high promontory, overlooking the entrance to the Bay of Naples and Capri on the southern horizon. By-passed through the centuries by the hordes who visit Capri, Ischia has more unspoiled natural beauty, if a somewhat less cosmopolitan atmosphere, than its sister island. Ischia's volcano, covered with shrubs and vines, helped to give it the name "Emerald Island."

Elba's shores, lying just a few miles off Italy, were first visited by the Etruscans about 700 B.C. The iron they sought here has never run out, but visitors arriving now are more attracted by the serene beauty of the island.

THE TYRRHENIAN SEA

THE BAY OF NAPLES

CAPRI and **ISCHIA** are the largest of several small islands that guard the Bay of Naples. Capri, small in size, is one of the world's most famous islands due to its reputation as a sun-drenched playground for the "jet set." It was, however, a tourist resort 2,000 years before jet aircraft. Often hailed as the world's most beautiful island, Capri was preferred by Tiberius Caesar even to Rome. Unwilling to leave his several villas on the island, he ruled the Empire by sending smoke signals to the mainland!

The Blue Grotto, a low, vaulted sea cave that is entered only by small flat boat, is the island's most renowned attraction. The grotto takes its name from the azure glow of sunlight reflecting through the water that almost fills its entrance. Other points of interest include the ruins of many great Roman villas. About 7,000 people live on the island.

Ischia, lying just across the mouth of the bay from Capri, has two dominant features—a pyramidal volcano called Monte Epomeo and a great medieval castle. The volcano is now inactive and the castle crumbling, but this island offers an excellent opportunity for discovering isolated vistas. The view of Naples and Mt. Vesuvius on clear days from the higher land is unmatched.

Thermal springs on Ischia are unusually high in radioactivity. Not as well known to tourists as its sister, Ischia depends upon fishing and wine production, rather than tourism, for its income.

THE TUSCAN ARCHIPELAGO

ELBA is the largest of the 7 small islands of the Tuscan Archipelago. It lies between Corsica and the Italian Peninsula about 8 miles from the mainland city of Pisa. The coastline of Elba is rocky and rugged and its landscape soars to more than 3,000 feet above sea level. Its 85 square miles have provided rich iron ore deposits since ancient times, and nearly one half of the island's 50,000 people still work in mining. Portoferraio with a population of 12,000 is the only town.

The Etruscans were the first to settle in this island, and later it was held by Greece, Rome, France, Spain, Genoa, Pisa, the Turks, and the Papacy. Napoleon was exiled here after his first abdication in 1814, before he returned to France and final defeat at Waterloo.

The islet of MONTE CRISTO, which lies a few miles south of Elba, was immortalized in Alexandre Dumas' novel, *The Count of Monte Cristo.* Only a few fishermen live on the island today. Other members of the Tuscan

Portoferraio is the only town on the island of Elba. It lies on a narrow coastal plain with green highlands rising sharply behind it. It is a typical Mediterranean town, with most of the houses painted pale orange and topped with red tile roofs. The port is home to a moderately active fishing fleet and receives launches from Pisa and other mainland ports.

Elba's history has been that of Tuscany, the part of the mainland nearest to the island. The Tuscan city-states of Pisa and Florence both occupied Elba and put its iron deposits to use. This stately fortress at Portoferraio is named for the Medici, Florence's greatest family.

The letter "N" is found on buildings throughout Elba, and stands for Napoleon. This is a museum that contains many of the Emperor's personal effects, writings, and other articles.

Archipelago are GORGONA, CAPRAIA, PIANOSA, GIGLIO, and GIANNUTRI.

Near the mainland coast, between Tuscany and Naples, are the PONTINE ISLANDS—PALMAROLA, PONZA, and VENTOTENE. During the time of the Caesars, these were popular places to exile political enemies or disloyal relatives.

ADRIATIC ISLANDS OF ITALY AND YUGOSLAVIA

During the Middle Ages the city-state of Venice ruled most of the islands in the Adriatic, that narrow finger of the Mediterranean that separates the Italian and Balkan peninsulas. Today, however, most of these islands are part of Yugoslavia.

The city of Venice, itself, is made up of many tiny islands at the head of the Adriatic. LIDO, long and sandy, facing open sea, is a fashionable resort area of the city of Venice. Nearby is MURANO, a village long famous for blown glass, built on five small islands. To the south is BURANO, where exquisite lace has been produced for centuries.

The northernmost island in the Adriatic, GRADO, belongs to Italy. The only other Italian islands, PIANOSA (there is another island with the same name in the Tuscan Archipelago) and the TREMITI group lie off the Italian coast,

A fisherman on Monte Cristo—"the Mountain of Christ"—weaves baskets of reeds for catching sardines. Though one of Italy's smallest populated islands, Monte Cristo is widely known because of Dumas' novel about a fictitious count who took his title from the name of the island.

Hvar is the name of the port, town and island that lure more European vacationers than any other part of the Adriatic. Luxury cruise ships often stop off here for a look at the "lavender island" on the way to Dubrovnik, mainland Yugoslavia's chief resort. Hvar lies among countless islands, each with its own ancient history, and is probably the best base for sailing through the Adriatic or searching for sunken Roman ships.

A group of tiny islets called the "Devil's Islands" lie off the town of Hvar. From the air, these dusty islands, surrounded by the blue-green Adriatic, form an amazing pattern resembling a work of abstract art.

A Venetian fortress looms over the waterfront of Korcula.

halfway between Pescara and Barletta, near the "spur" of the Italian boot.

The Yugoslav islands, most of which are long, thin slivers lying near the Balkan coast, are far more numerous—there are hundreds of them, many so small they have never been named. KRK, CRES, DUGI, OTOK, VIS, PAG, BRAC, HVAR, KORCULA are among the largest.

Their historical ties to Italy are evidenced by impressive Roman ruins and beautifully preserved Venetian villages. Marco Polo was born on KORCULA, near the mainland port of Dubrovnik. Ten miles away, HVAR has become the most popular island among tourists because of its unique beauty. Fields of lavender cover the island hills, and forests of pine and poplar grow down to the high-tide mark where occasional palms are found. KRK, the most northerly of Yugoslavia's islands, enjoys some of the best fishing in the Mediterranean. Especially charming is the village of Baska, with cobblestone trails that wind up the steep inclines. The island of RAB, south of Krk, has remarkably preserved Roman roads and medieval towns as well as some of the best modern hotels found anywhere in the Adriatic.

The Yugoslav island of Korcula has a Venetian atmosphere. Marco Polo, who opened the Orient to Europe, was born on the island, probably in the house with a lookout tower, in the middle background.

A blend of Renaissance, Arab, and other styles, gives Valletta, Malta's capital, a special look. The Dghajsa boats shown here were first introduced by the Phoenicians and still ply the broad waters around the city as water-taxis. During World War II, Valletta and Malta avoided occupation during three years of heavy bombing by the Axis powers. For their gallant defence, the British Government awarded the George Cross to the entire Maltese population.

MALTA

A GEOGRAPHER'S sketch of Malta could easily lead one to believe that this independent island country at the very heart of the Mediterranean is uninhabited. There are no lakes, rivers, or mountains and the soil is of the poorest sort and extremely thin. But this is not the true story of Malta. The former British colony is, in fact, one of the world's most densely populated countries, with more than 300,000 people living on 122 square miles.

THE LAND

The archipelago that makes up Malta includes three small islands and two others so tiny that they have no inhabitants and seldom appear on a map. The island named MALTA is the largest and most important, measuring about 22 miles in length and 8 miles across at its widest point. Lying to the west of Malta, GOZO and COMINO are the smaller islands, the latter being barely a mile wide.

Located about halfway between Italy and the coast of North Africa, Malta occupies a position that has been of great commercial and military importance throughout European history.

Ages ago, a bridge of land connected Italy with Africa, dividing the Mediterranean into two great lakes. The islands of Malta are a fragment of that bridge.

As the waters of the sea rose around Malta they crept into its valleys, making some of the finest natural ports in the world. Two of the deepest and best protected of these flank the capital city of Valletta.

The only fresh water on Malta comes from deep wells. Beneath the thin soil is found the sandstone, Malta's only mineral resource, used for most of the buildings on the island.

Besides serving as the country's political capital, Valletta, which looks toward Sicily to the north, is also the home of more than half of the Maltese people living within its environs. Valletta, seen from the sea, resembles a

With no rivers or lakes, the Maltese have to rely upon deep wells for their supply of fresh water. Old windmills, like this one at the seacoast village of Wied-iz-Zurrieq, put the prevailing Mediterranean winds to work, bringing the water to the surface. The windmill, buildings, and walls seen here are all made of the native limestone.

monumental fortress. The city is surrounded by a very deep moat and some of the strongest fortifications in the world. They were constructed during the 1500's to keep the Turks out, and today their awesome beauty attracts visitors from all over the world. Behind the great walls lies the Mediterranean's most beautiful and best planned Renaissance city.

Across the island from Valletta, the more ancient city of Mdina rises like a mirage above the rolling plain. St. Paul paid a visit here when he was shipwrecked on Malta. Most of the buildings date from medieval times, while in caves below lie the undisturbed fossils of prehistoric animals, including elephants and a giant rodent.

The many tiny coves that pierce the coastline of Malta provide feeding grounds for fish and a way of life for the fisherman. On the smaller islands of Gozo and Comino especially, fishing is the main occupation.

At Paola, a village at the head of one of Valletta's ports, is the Hypogeum, a 4,400-year-old temple, with artifacts and altars in good condition. The prehistoric ruins on Malta, Sardinia and other islands of the Western Mediterranean were long thought to have been built under the influence of the Minoan civilization. Recent carbon-dating techniques show that these and other western European ruins are older than any structure in Greece or Egypt.

While street decorations may indicate a holy day on other Mediterranean islands, they are part of the landscape on Malta, where every day is a religious celebration. A typical street in a Maltese town is lined with statues of patron saints, shrines, chapels, and churches. Even small villages—some with old Arab names—often have huge Gothic or Romanesque cathedrals. There is no place in the Mediterranean where religion—and on Malta that means Roman Catholicism—is more a part of daily life.

The main town on Gozo is Victoria, named by the British. Of special interest here are ancient windmills, still turned by the persistent trade winds known in the Maltese language as *xlokk* and *nofsinhar*. A hilltop citadel at Victoria dates from the 17th century. Calypso's cave, where, according to legend, Odysseus lived for a time, lies on Gozo's northern shore.

Comino is too small for towns. An outcropping of rock with almost no soil, the island is blanketed with thyme and populated by fishermen. Hundreds of tiny hidden coves make it a popular stop for boaters.

HISTORY

The written history of Malta begins with the coming of the Phoenicians about 600 B.C. The country later fell under the domination of Carthage and was taken by Rome during the Punic Wars. When the Roman Empire was split, Malta went to Constantinople (Byzantium). The Normans and the Spanish Crown each held the islands for a while. In 1530, Malta's greatest era began when the Holy Roman Emperor presented it to the Knights of St. John of Jerusalem. These Knights had

accompanied the Crusaders to the Holy Land, settled on Rhodes, lost to the Turks and needed a base of operation. The Knights successfully defended Malta against the Turks, who were ravaging many Mediterranean islands, and built the island's most enduring structures.

Through the years the Knights became more and more French, as did Malta until Napoleon finally claimed ownership in 1797. The British captured the island in 1800 and remained until Malta became independent in 1964. During the years between 1953 and independence, the North Atlantic Treaty Organization (NATO) maintained its supreme Mediterranean headquarters there, staffed mainly by the British Navy.

THE PEOPLE

The purest descendants of the Phoenicians may very well be found on Malta. Their language, Maltese, may be partly derived from Phoenician, although this is a matter of dispute, for others claim that Maltese is an offshoot of Arabic. It is agreed, however, that Maltese is a Semitic language, like Arabic, Hebrew and the dead language of the Phoenicians. The Maltese are thus the only Europeans who speak a Semitic language!

This stone-paved plaza lies in front of Valletta's Church of St. Publius. The large stone slabs, boulders cover 16th-century subterranean silos where the island's grain is stored after harvesting.

In Malta almost everything is built of big limestone blocks cut from the island's bedrock. The older limestone buildings have turned a rich golden shade, while new structures shine brilliantly white in the sun.

The Miziep, a stretch of rolling hills on Malta, is one of the few areas in the country where the soil is fertile and deep enough for sustained farming. This farmer and his son are picking pumpkins.

Maltese and English are both official languages, and Italian is widely understood here.

The Maltese are an intensely religious people and Roman Catholicism, the state church, pervades their lives. Most celebrations in the country are religious in nature. Every town and village has a church that serves also as the social headquarters of the people. The church at Mosta, a town of 8,000 people, has the world's third largest dome.

THE ECONOMY

Malta is an agricultural country, but because of the density of population, it is impossible to grow enough food for the people. Most of the foreign trade is still with Britain.

Since the British navy left, the Maltese have successfully turned to tourism as a means of boosting income. A few small industries produce furniture, office equipment, clothing—including American-type jeans, a popular item—and fiberglass boats for export.

A new campus for the Royal University of Malta opened early in 1971. When all the buildings are completed, over 650 students will be accommodated. The university was founded as a Jesuit school in 1592 and became a university in 1769. Even here, where the architecture is ultra-modern, the traditional limestone blocks are used.

Corte, in the central uplands, is the historic old capital of Corsica. The town clings to a peak rising abruptly from a bare plateau. Its fortress was begun in the 15th century. Houses here are made of schist dug from the mountain just as they were during the long struggles against the Genoese occupation.

CORSICA

CORSICA HAS a way of reaching out to welcome the visitor approaching by sea. The sweet aroma of its lush shrubbery—rosemary, honeysuckle, thyme, myrtle, and lavender—drifts in the breeze across crystal-clear waters, offering just a hint of the beauty that is waiting on the Mediterranean's "Scented Island."

Terraced mountain slopes of dark green and little houses with whitewashed walls and red tile roofs present a visual display that matches the sweetness of nature's perfume. Corsica is much the same today as it was almost 200 years ago when the island's most famous son, Napoleon Bonaparte, left for the mainland, destined to become Emperor of the French.

Though they are arriving in increasing numbers, tourists have not greatly altered Corsica's beauty or way of life. Both Corsicans and visitors seem to have resolved to keep Napoleon's Corsica as he would have remembered it.

THE LAND

The Mediterranean's third largest island, Corsica is shaped roughly like an oval, lying off the Riviera, about 100 miles from Nice, 50 miles from Italy, and within sight of Sardinia, which lies across the narrow Strait of Bonifacio on the south.

The dominating feature of Corsica's landscape is its towering mountains that rise to more than 9,000 feet above the sea. Monte Cinto, the highest peak, is only 15 miles from the sea and

The town of Porto overlooks a beautiful, narrow, fjord-like gulf. Because of the grandeur of the scenery many tourists visit this town. Here, donkeys ridden to town by farmers graze beside a cosmopolitan outdoor café.

other mountains of over 7,000 feet lie so near the water's edge, that they rise like great walls holding back the tides. The mountains are often called the Corsican Alps, and, in fact, villages at higher elevations experience a bracing, almost Alpine climate, with snow on the ground more than half the year. All the big towns and cities are on the lower plains.

Herds of moufflon, a kind of wild sheep, inhabit the rocky highlands. To make more land available for farming, many slopes are terraced like mammoth stairways. Much of the

The greatest number of Corsican villages lie on the coast or along gulfs, with well-protected ports. Almost every house on the island has a roof of red tile.

The narrow, twisting lanes of Bastia offer an explosion of sights, sounds, and smells to tempt the senses. Young Corsicans today grow up in an environment not greatly changed since the time Napoleon was a lad on the island.

land not under cultivation is covered by the thick growth that provides the island's special scent—collectively these plants are called the *maquis*. At places the *maquis* grows so thick it has traditionally been the chosen place of bandits from the mainland. During World War II, the word *maquis* came to stand for the entire French underground fighting the Nazis. There are some forests of chestnut, oak, and beech on the intermediate slopes.

The three most important cities of Corsica are all ports. Ajaccio is not the largest city, but is by far the most important in the minds and hearts of all Frenchmen. It was here that Napoleon was born on August 15, 1769. Ajaccio lies on the west coast of the island beside a sheltered gulf. The capital of Corsica—the island is a department or province of the French Republic—Ajaccio has more than 40,000 residents, but retains the charm of many small fishing villages.

The largest city on Corsica, and the nearest thing to a metropolis, is Bastia, population 50,800. A mixture of modern apartment

Ajaccio is situated at the head of a gulf, enclosed by noble mountain ranges that shelter it from prevailing winds. Here is found the most ideal climate on Corsica and the greatest number of tourists. At the quay is a British tourist ship and modern hotels can be seen on the hills above the old section of town.

houses blends with traditional Corsican architecture behind the quays where boats from Nice dock. When Genoa ruled Corsica, this was the capital.

A gift of the 13th century, Bonifacio, perched on a narrow promontory facing Sardinia, is one of the most unusual towns in Europe. With its massive walls and ramparts, and underground reservoirs fed by rain water, it was able to resist long sieges in the Middle Ages. One still enters the city via the same antique drawbridge and huge wooden doors which Charles V, the Holy Roman Emperor, used when he visited here.

HISTORY

Though human history on the island dates to the Stone Age, and Corsica rode the sometimes rough waves of each epoch of Mediterranean civilization, the islanders see Napoleon's birth as the culmination of Corsican history.

Born of Italian parents just after the end of Genoese rule, Napoleon once supported war with France for Corsica's independence. Later he had a change of heart, learned to speak French—only Italian was spoken in Ajaccio during his youth—and left for Paris to join the army. Years later, after his final exile to St. Helena, Napoleon, the fallen leader, seems to have had regrets that he gave Corsica so little attention while he was Emperor. "Oh, Corsica," he wrote. "I did nothing for Corsica. All I did was have a fountain built."

Corsica's earliest monuments predate Napoleon's fountain by at least 3,000 years. On the southwest slopes of the island, large stone carvings, resembling those found on Easter Island in the Pacific Ocean and called

To make more land available for farming, many of the island's slopes are terraced with retaining walls of rock. Not only does this arrangement give the farmer more land to work with, but it also prevents erosion. Grapes for wine production are often grown on such hillside farms.

Napoleon's birthplace, on a sleepy street in Ajaccio, is the island's most important landmark. Here, on August 15, 1769, the future Emperor of France was born, exactly one year after the French took the island from Genoa. The town is still full of reminders of him, and the house is filled with mementos from his childhood and career.

menhirs, face the sea with frozen, intense faces. They are the legacy of a megalithic civilization (using giant stones) that was Corsica's first.

The Greeks arrived in 560 B.C. and founded a town known as Aleria on the east coast. Etruscans and Carthaginians vied with Greece for control of Corsica, but it was the Romans who finally forced the Greeks off in 250 B.C. Rome made good use of the entire island, building great cities and protected seaports for its merchant fleet. Christianity was brought to Corsica by the Romans during the 3rd century, though the oldest churches on the island date from the Middle Ages.

The region around Sartene is regarded by many as the most characteristic part of Corsica. Powerful feudal lords once ruled the area, and here medieval resistance to the Genoese was longest and fiercest.

The music, dance, and tradition of the rural Corsican are a blend of French and Italian influence, with a little of its very own. Religious festivals, such as Easter, are always occasions for celebration, but no holiday compares with Napoleon's birthday for excitement on Corsica.

After the decline of Rome, Vandals, Ostrogoths and Moors invaded Corsica, destroying most of the island's classical architecture. The Eastern (Byzantine) Empire ruled for a while until Charlemagne, Emperor of the Franks, returned the island to the West. Finally the Roman Catholic Church assumed ownership in 1077 and Pope Gregory VII gave the island to the Bishop of Pisa.

Genoa, a rival city-state, objected to Pisa's control and won half of Corsica from the church in 1133 and the other half in battle with Pisa in 1284. Genoa was the supreme—though not unchallenged—power on Corsica until 1768—the year before Napoleon's birth, when the island was sold to France. The following year Louis XV made Corsica a province of France—a distinction no other ruler had granted the island. Since then the history of Corsica has been that of France.

The roads of Corsica are crude and sometimes dangerous for any mode of transportation but the sure-footed donkey. Kegs strapped to the side of the animal are used to carry goat's milk into town.

Like the mainland, Corsica was invaded by the Germans during World War II, but the determined Corsicans banded together and liberated themselves a year before the invasion

Sheep are commonly raised in the mountainous regions of Corsica. Their milk is used for making cheese, particularly a kind resembling the famous Roquefort of mainland France. These sheep are grazing along the shores of Lake Nino.

The streets leading to the old Genoese citadel above Calvi are narrow, winding, and paved with rough stone—more suitable for walking and riding donkeys than for motor traffic. In many of the smaller towns the traditional dress is still worn.

of Normandy—the Germans simply could not get at them in their mountains and their *maquis.*

THE PEOPLE

Most of the 280,000 Corsicans live a simple life, unmoved by the tensions and troubles of the mainland. They have a tradition of silence and aloofness, but in their own way enjoy life intensely. Fond of bright hues, the Corsicans can be found on their beaches and winding mountain roads on weekend outings, enjoying the vivid landscapes of their countryside. During the week the village markets are usually alive with activity.

The French spoken on the island is heavy with borrowed Italian phrases and accents. The native speech, a form of Italian, is spoken in the hinterland and villages.

On small farms, grains, fruits, and vegetables are grown for the farmer's family. Goats supply milk and cheese. If a Corsican is not a farmer, chances are he fishes for sardines or shellfish for a living.

THE ECONOMY

Because there are few jobs available on the island for people who don't care for farming or fishing, many Corsicans have followed Napoleon's example and left for Paris and other mainland cities. Except for a hydro-electric plant, there has been very little recent industrial development on Corsica. French President Georges Pompidou has pledged that the government will try to do more to develop the island Department.

Corsica exports the same goods it has been known for since merchant ships first started calling at its ports. Olive oil, timber, wine, cork, and a sharp white cheese similar to Roquefort are important. At Canari, there is an asbestos quarry, but no other mineral wealth has been found. Most of Corsica's products go to the French mainland.

Tourism is the fastest growing business on the island, with August being the busiest month for hotels and open-air cafés. Most of Corsica's tourists come from Paris.

59

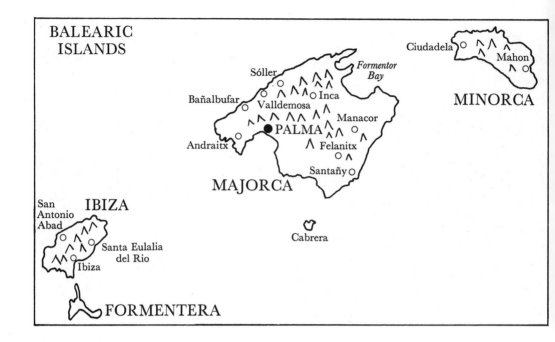

BALEARIC ISLANDS

Ciudadela
Mahon
MINORCA

Sóller
Formentor Bay
Bañalbufar
Inca
Valldemosa
Manacor
PALMA
Andraitx
Felanitx
Santañy
MAJORCA

Cabrera

San Antonio Abad
IBIZA
Santa Eulalia del Rio
Ibiza

FORMENTERA

THE BALEARICS

REACHING TOWARD the open Mediterranean from the ancient coast of Iberia like a beckoning salute to the east, the Balearics are appropriately known as the "Isles of Peace." Through the ages they have offered hospitable shores to each civilization that touched the mother country, Spain, not as island outposts, but rather as hubs of trade and culture. Their great natural beauty inspired Chopin's sprightly preludes and today the islands are just about the most popular destinations for visitors to Spain.

GEOGRAPHY

Fifteen islands and islets make up the Balearics and the Spanish province of Baleares. Mallorca (its English name is Majorca) is the largest and best known of them, with Menorca (Minorca in English), Ibiza, and Formentera completing the list of major islands. Some of the islets, mostly grouped around Ibiza in the south, are just a few square miles in area.

All the islands are the tops of submerged mountains, an extension of the rugged ranges of southern Spain, and stretch from southwest to northeast. Minorca, the northernmost island, lies about 120 miles off Barcelona, while the southern islands are no more than 50 miles from the mainland province of Alicante and 80 miles from the port of Valencia.

The Balearics have the customary Mediterranean climate with perhaps somewhat less rain than the average. The soil is dark and fertile, though, and with irrigation systems that date to the Roman era, the islanders have added a rich blanket of green to the montage of golden beaches and blue sea.

Collectively, the islands have almost 2,000 square miles of land, more than half of which belongs to Majorca, and a population nearing 500,000.

MAJORCA

Majorca is the hub of everything in the Balearics. Its 47 by 62 miles contain the greatest

The port of Palma has provided shelter for great merchant ships and the smallest fishing boats since man first began to sail the Mediterranean. Here a lone fisherman prepares to go after a day's catch.

population, the most extensive agricultural, industrial and tourist development, and the provincial capital of Palma. It is also situated in the middle of the archipelago, with Minorca lying 20 miles northeast and Ibiza 50 miles southwest.

Formentor, the most remote area on the island of Majorca, is the most beautiful recreation ground in the Balearics. Sandy beaches and wooded coves with crystal-clear water draw thousands who are interested in water sports.

Two mountain ranges flank opposite sides of the squarish island, with a rich, level plain in between. The western range is high and rocky, while the eastern hills are lower and rolling and contain some of Europe's most impressive stalactite caves.

Palma, the capital and largest city of the islands, is also the chief commercial city and has been since ancient times. It lies on a magnificent circular bay that cuts into the southwestern coastline of the island. Casting a golden reflection into the glass-like waters of the bay, Palma is a city of all ages. The narrow streets follow winding routes laid out by the Phoenician, Greek, Carthaginian, and Roman colonialists, past buildings that represent every era from the 12th century to the 1970's. The 800-year-old cathedral, second largest in Spain, seems to sit precariously near the sea. Two palaces loom above the city as constant reminders of its Moorish past. The Almudaina Palace served as the seat of Majorcan kings during a short period of independence in the Middle Ages, and Bellver Castle is a towering fortress surrounded by two moats. In an old convent a few miles outside Palma, Frederic Chopin spent a winter with the French novelist George Sand. The cell where he composed and his piano are now on exhibit there.

Other points of interest on Majorca include Manacor on the eastern shore, headquarters of the island's thriving artificial pearl industry, and Inca, an inland village in the north known for its fine leather goods. Formentor, on the opposite side of the island from Palma, has the most remote and beautiful of Majorca's hundreds of beaches, and along the northern coast, on the Bay of Alcudia, are extensive Roman walls and ruins, but no modern towns.

MINORCA

The second in size of the Balearics has barely half the area of Majorca—35 miles long and 10 miles wide—and is much flatter. For centuries, Minorca's farmers have been picking up the rocks and boulders from their fields and building walls that fan out into intricate patterns that give the entire island the appearance of a jigsaw puzzle. In almost every field, an old windmill slowly turns, giving water to the crops and atmosphere to the rolling landscape. It reminds readers of *Don Quixote* of the windmills of La Mancha in mainland Spain.

The main town on Minorca is Mahón, founded by the Carthaginians as Portus Magonis, after Mago, the brother of Hannibal. The old town rises in terraces from the best natural port in the Mediterranean. Besides excellent relics of its ancient beginnings, Mahón is known as a cosmopolitan city, long popular with the French and British. Among the city's contributions to Western culture is mayonnaise (formerly *mahonnaise,* in French).

On the opposite coast of Minorca, facing Barcelona, is the Moorish city of Ciudadela, with ornate palaces and a Gothic cathedral.

Scattered across the island are more than 500 stone mausoleums dating from the Bronze Age. The Minorcans contend they were constructed by a race of giants, and indeed archeologists are baffled as to how prehistoric men moved such enormous stones.

IBIZA

Ibiza is the most westward of the Balearics. About 40,000 people, including a large community of British and American expatriates, live on this little sun-baked island, just 25 miles long and 8 miles wide. The moorish influence is more predominant here, each tiny village having

The Spanish Government has opened many convenient camp grounds in the Balearics, such as this one near the coast on Ibiza. Occupying a wooded area, it is within walking distance from several secluded coves, sandy beaches, and archeological sites. In the cities, hotel rates tend to be quite moderate and the tourist season lasts just about all the year.

at least a few old Moorish buildings still in use. The only sizeable town is also called Ibiza and dates from 654 B.C. when Carthage ruled. The Romans maintained Ibiza as an important port, but it was not until the 13th century that the present city began to take shape. In 1585 a wall was completed that encircled the entire city, protecting its citizens from the terror of Barbary pirates. San Antonio Abad, 8 miles west of Ibiza, is the island's most popular tourist attraction, with long palm-lined beaches and clear warm surf. Santa Eulalia del Rio, just north of Ibiza, is appropriately named, as it sits on the only river in the Balearics.

FORMENTERA

Formentera, at the extreme southern end of the chain, and 11 other tiny islands in the area, have no important towns. Fishing is the main source of income. Lately tourists have discovered these remote areas that offer

Gothic, Romanesque, and Moorish architectural styles blend delightfully in Palma's Tower of Santa Catalina of Valencia. Palma has not maintained the character of any one era of its history, but draws from all of them.

With the fall of the Roman Empire the Vandals moved into Spain and occupied the islands. Their rule was later overthrown by forces of the Byzantine Empire.

It is unlikely that anybody on the islands took note when Mohammed was born in the 6th century A.D., but 300 years later the Balearics were taken by his followers in the western Mediterranean, the Moors.

For centuries the kings of Europe, especially those of northern Spain, tried vainly to win the "Isles of Peace" back to Christianity. But it was 1220 before any of them succeeded. In that year the Balearics came under the control of Aragon, first as a semi-independent kingdom— there were four Kings of Majorca—then as part of Aragon. With the exception of Minorca, they have remained Spanish ever since.

Minorca was taken by the British in 1756 at the same time that they acquired Gibraltar. Later it fell to the French, then back to the British, who finally gave the island to Spain in 1802.

excellent unspoiled beaches and rocky coasts perfect for skin-diving.

HISTORY OF THE BALEARICS

Iberians, the same forgotten race that first civilized the mainland, were the original inhabitants of the Balearic islands. During the centuries before Christ, Phoenicians, Greeks, and Carthaginians in turn built fine cities and temples on the islands and trade flourished regardless of which civilization supplied the merchants.

The Carthaginians left the city of Mahón as their most enduring legacy, while the Romans bequeathed Palma.

THE PEOPLE

Despite the fact that the Balearics have a history full of foreign invasions and landlord governments, the people remain essentially Spanish in traditions. Their language, Catalan,

Fishermen, in their traditional black berets, fold handmade nets for storage after a day at sea. Scenes like this have occurred daily for centuries, but the background is something new to the islands—a modern luxury hotel.

63

THE ECONOMY

Farming and fishing are the main occupations of the Balearics—just as on most Mediterranean islands. Olive groves cover much of the land and the oil, pressed by windmills, is a major export product. Also grown are figs, almonds, apricots, lemons, oranges, grains, pimientos, and grapes.

Some fine wines are made on the islands, as is a crumbly brown cheese called *queso mahones,* which is becoming popular throughout Europe.

Most of the fish caught by Balearic fishermen are eaten on the islands. Some fishermen have found it more profitable to convert their fishing boats into passenger boats for ferrying tourists among the islands and to and from the mainland.

The annual influx of tourists is now more than 3,000,000—six times the native population —and growing each year. The money which foreigners spend in the Balearics is an important part of the islands' income.

shows a resemblance to French, but can be understood by Spanish speaking persons. Catalan is also the language of Andorra and of northeastern Spain.

Life on the islands proceeds at an easy pace. Getting water to the fields and olive groves and harvesting the crops occupies the farmers' time. Many farmers live in small towns and drive their families several miles to the fields each morning for a hard day's work. The big cities offer a more cosmopolitan existence, where local citizens are often outnumbered by visitors.

On major Catholic holidays, however, people in the city and those in the countryside see eye to eye. A festival in the Balearics calls for guitars and ancient bagpipes, and parades with paper horses. The islanders have many traditional songs for such occasions.

The unique Bellver Castle, outside Palma, was once the royal home of a King of Majorca and a prison for another. Built during the 14th century, it is a curious combination of circular towers, courtyards and Gothic arches. Although it was begun by the Moors, it does not follow the usual Moorish lines.